Science of the Day
versus
Everlasting Science

D.G. Boland, Ph.D.

En Route Books and Media, LLC
Saint Louis, MO USA

En Route Books and Media, LLC
5705 Rhodes Avenue
St. Louis, MO 63109

Contact us at contactus@enroutebooksandmedia.com

Cover art by Sebastian Mahfood

Copyright 2025 D.G. Boland

ISBN: 979-8-88870-364-9

All rights reserved. No part of this book may be reproduced, stored in a retrieval system, or transmitted in any form, or by any means, electronic, mechanical, photocopying, or otherwise, without the prior written permission of the author.

Table of Contents

Preface iii

Science of the Day versus Everlasting Science.......... 1

Preface

The present book deals with the difference between the meaning that science has for the modern mind, which has come to be known as modern science, from that which it had and has in Aristotle and Aquinas. which historically is known as Aristotelian. I have already dealt with the profound difference in my book "Teenager Science and the Curse of the Creative Imagination".

There is a difference in treatment between what I have brought out of the difference and opposition in my earlier treatment and what I point to here. Both treatments are to do with ways in which modern science differs fundamentally from the meaning of science which had been accepted as classical since Aristotle.

In the previous treatment I focused on how modern science openly and directly rejected and rejects the Aristotelian notion of science. This is commonly recognised as excluding Metaphysics from any notion of science and even regarding such an intellectual exercise as anti-scientific. I also pointed there to the extension of this exclusion of the meaning of science to any explanations by way of natural formal and final causes within Natural Science or Physics. The reader may find this fully set out in my previous book.

In the present book I wish to bring out the difference in a way that is more subtle and indirect. The fundamental opposition is still there, put not in explicitly negative terms, but by presenting modern science in positive terms in which the opposition is implicit and as it were hidden. For there is nothing that is purely and entirely bad or wrong and one can always accent the good and true that exists still in what is evil or erroneous.

Many are taken in by this clever (and even diabolical) tactic. And those who wish to argue in favour of modern science, even where it is clear that it wrongly denies the scientific value of metaphysical and natural philosophical principles, make great use of this sophistical tactic. The way this is done is to focus on limiting science to the study of those aspects of reality that are the objects of physics and mathematics, and in particular as presented in the mixed science of physico-mathematics in the manner we have explained in our previous books.

Here the main objects of investigation are quantitative aspects of physical reality, which as Aristotle noted, is defined in terms of motion and time. Though there is much left out in the investigations of modern science, there remains a residue that is amenable to closer study and analysis that can produce results of scientific truth and value. The deficiency of this approach and "scientific method", if partial and deficient, as explained inevitably produces the many evils both of falsity in theoretical subjects and of distortion and disorder in practical matters we have pointed to in the present book.

This tactic of accentuating the positive and ignoring the negative, is widely used and is so influential in compulsory education in modern science, including many a Catholic institution, is what we focus on in the present book.

The two ways of seeing the opposition between Aristotelian and modern science though related are distinct. This book is therefore put out as a stand alone one. However, it ought to be read in conjunction with "Teenager Science and the Curse of the creative Imagination". For this reason, we do not have a separate dedication and introduction.

Science of the Day versus Everlasting Science

To further clarify the meaning of Science we will first treat Aristotle's definition and then contrast the modern notion with it. The classical definition of Aristotle adopted by most and especially by Aquinas is quite straightforward and fulfils perfectly the conditions of definition in Logic as explained by St. Thomas. Let us first state the definition, which can be seen to be in two parts

certain knowledge (genus)
through causes. (specific difference)

It will be noted that this definition applies to all sciences, even though we make a distinction between the kinds of causes that apply to different sciences (Mathematics for instance uses only intrinsic causes). It might be noted here that science is a habit of intellectual knowledge. Another habit is called understanding, and it is distinguished from science in that it does not know through the medium of causes (or reasoning) but is certain knowledge that is immediately known. Such is the knowledge of first principles. These belong to the second act of the mind, judgment, whilst science as a habit belongs to the third act. It would seem that intellectual habits need to be based on the act of judgment which is a perfect act. Concepts which are a product of the first act of the mind, simple apprehension, only give us an imperfect knowledge of things.

There are of course many acts of intellect that fall short of certainty so they differ generically from science. Dialectic, for instance, only gives us knowledge of things according to probability or opinion.

What may be noted here is that there is a most radical difference between the Aristotelian and modern (Newtonian) notions (and definitions/descriptions). For modern science eschews any claim to certain intellectual knowledge (at best seen as only a dialectical exercise) and even goes so far as to proclaim radical scepticism in regard to all scientific knowledge. It virtually reduces all human knowledge to sense knowledge, and thus human nature to the level of brute animals. But we will come to that shortly.

As noted in our book "Teenager Science and the Curse of the Creative Imagination", there are basically six orders of science. Of these St. Thomas differentiates one, Logic, as instrumental, and five as principal. Logic has features that are peculiar to it. Though able to be divided into formal and material aspects, it is primarily focused on the formal part of the concept, proposition and argumentation. Its material aspect is real essences of things but taken only generally. Taken specially, the material aspect relates to things outside the mind.

However, it should not be forgotten that all things in mind involve logical relations, such as universality added to the concept horse. This follows necessarily in the mind from the things known being abstracted from material conditions. Dialectic is peculiar in that the material part of the object of study is not reality but what has logical status only.

The five principal sciences are differentiated into three theoretical (for the sake of knowing only) and two practical/productive (for the sake of doing/making). The three theoretical sciences are Physics, Mathematics and Metaphysics. They all involve the three acts of the mind ending in reasoning. Each act of the mind, simple apprehension, judgment and reasoning has formal and material aspects, though this combination has to be carefully understood in each case. The focus of the principal sciences is on the material aspect (such as the abstract essence of horse in the concept) but it also contains the formal aspect (the logical relation of universality in the concept of horse). Similar analyses apply to the practical/productive sciences.

But we leave the study of the details to our previous books. What is especially important to note is that the object of Mathematics is something real in the physical or bodily order. It is just that the abstraction involved leaves out every accident belonging to material things except quantity. The modern mind has all sorts of difficulties in determining the status of the object of Mathematics, some putting it with logical relations, others with a separate ideal or meta-physical order. Even the ancients and medievals were not able to get a clear idea of the difference between logic, metaphysics and mathematics. Plato got things mixed up here, especially not able to clearly distinguish Mathematics from Metaphysics. It was not until Aristotle that the distinctions between the three theoretical sciences, and also the practical/productive sciences were sorted out. Even there it is not easy to follow Aristotle without Aquinas.

All this becomes messier when we bring in the mixed science of physico-mathematics or mathematical physics. Aristotle was quite aware of this as a separate and distinct order of theoretical science. But he did not develop this mixed science to any great extent. He noted such particular examples of this order of sciences in perspective, music and astronomy. But the full understanding of the general notion of the mix of mathematics and natural science remained elementary.

What needs to be noted in this regard is that physics, as Aquinas stated, is the material part of the science, while mathematics, the formal part, has to be considered as retaining its natural forms and ends. A significant feature of modern science is its rejection of natural forms and ends.

Thus there is a fourth order of theoretical sciences, known to Aristotle (and of great significance in seeing how modern science is understood quite differently from Aristotle and Aquinas). That is the mixed or medial science that has come to be called

Physico-mathematics. Indeed, taken generally it has come to be regarded as identified with the notion of science itself. However, there is a great difference between how Aristotle and Aquinas understood and applied this notion of a mixed science and the corresponding scientific method. The difference between Aristotelian and modern (Newtonian) science is discussed in our books and needs to be carefully studied.

The importance of this may be seen in the arts of painting and music. The quantitative forms and relations added by human creative imagination do not supplant the natural forms. The retention of the natural forms enables the painting or music to remain natural with its natural beauty and harmony coming through despite the addition of an artificial form from human art. One thing to notice in modern art and music is that the artists feel free to play around with the material elements (shapes and sounds) in their rawest material condition as it were. This can result in the production of monstrous distortions and dissonance/cacophony coming to be regarded as an advance of "modern art". This freedom to see nature denuded of its natural forms and merely as an empirical "mass" has much greater significance as we shall see when we come to consider the notion of science that came to supplant Aristotle's and to be called "Modern Science".

However, so far we can see that even science and art as classically set down by Aristotle has its complexity and it is by no means easy to provide a full and deep understanding. We can say though that St. Thomas has come as close as anyone to doing so. But even here we have to try to discern his thought presented in almost an incidental way as he inserted his insights in answering questions of theology in which he was primarily interested.

We will move now onto the "definition" of modern science as it developed from Isaac Newton, having made a break with what was considered to be the notion of science derived from Aristotle deficient in its neglect of empirical/material support. We need to make some comment in this regard, but it may be already appreciated that the mixed science of Physico-Mathematics generalized has a central place in the modern notion.

The word "modern' if anyone cares to look it up comes from the Latin *modo* which was the way the Romans said "now" or "of the day". When the self-styled modern world sought for a word to describe the science it wished to replace the science understood since the scientist who Dante called *il maestro di coloro che sanno* (the master of those who know), still having some faint grasp of Latin, it could not come up with any other name than "modern science". The problem with that name, as it became obvious, is that it refers to knowledge that lasts only as long as a day. For the same word is used for what is in fashion and appeals mainly only because it is new - satisfying the desire as Pope Leo XIII put it in his famous encyclical of 1891 on the miserable human condition of the majority (workers) *rerum novarum ... cupidine.* So it is that modern scientists are ever on the (re) search for something new digging into the depths of the material vestiges to be found in the natural world as we know it "to date". The modern scientists' one driving desire is for novelty in science (so that he can give his name or that of another modern man to the new scientific law ("fact") discovered, just as modern man (and woman) has an eternal itch for the fashion of the day, which using the Greek derived word is ephemeral, and disappears the next day. That is the essence of modern science from its very name. But so dominated is man by pride in his own superiority over everything else (especially his creator) that he flaunts his discovery of something new as the acme of scientific knowledge. With this goes of course disdain and even contempt for what belongs to the past, and to the timeless truths already uncovered by such genial minds as Aristotle and Aquinas. They are after all but men of yesterday in the language of the modern mind. The *gloria mundi* rests on the scientists of the present (for a day anyway). It does not take long for the days to pass (*sic transit*) and a new science is celebrated and taught to a young student body ignorant of practically everything worth knowing.

One may give a pseudo-definition of modern science, using cause in a mangled way, in terms of Aristotle's four causes. What I have given is the final cause, which is the end of the irrational appetite for the new. The notion can be refined further by giving the other three causes according to their pseudo-notions. That way we can focus on modern sciences formal object and method.

The quasi formal cause is the formal part of the mixed science of mathematical physics, as a combination of observations (practically reduced to mere sense recordings) worked upon to construct quantitative forms or formulas by the human imagination (feverishly devising "creative" new formulas). The mark of this feature of modern science is the mechanical nature of its explanations. It has some usefulness at the lowest material level but necessarily distorts any understanding of higher natural forms such as of living and sense knowing things. The most damage is done in dealing with human nature at the highest animal level, as in Psychiatry. Modern medicine is reduced to treating the human body as a machine. We should note that often this mathematico-hypothetical "creation" of theories is taken as the main feature of modern science. It marks the belief that the modern scientist has supplanted Aristotelian Metaphysics or "Philosophy" by Mathematics.

The chase for new mathematical formulas tends to go round in circles. However, diving more and more into the interstices of matter has its usefulness. Hence, the great success of modern medical investigations into the working of the human body (and soul at animal level) which depends on the integrity of the function of bodily elements.

Thus we can use in the pseudo-definition of modern science the material cause evident in the empirical and experimental character of modern science. As we have explained, this is where modern science has recovered what had been lost in the early Christian era after Aristotle. In a way we can say that Physics or the Natural Sciences became too "spiritual", as well noted by Josef Pieper. The advantage modern science had then over the pre-modern was played to great effect, resulting in the effective rejection of all Aristotelian work in the study of nature, including human nature.

This is where the devil has tricked modern thinkers most. But what we should also note is that the retrieval of the empirical method was not without its own distortion. Given the influence of the change, not to Physics as pure natural science, but to that mixed with Mathematics, we had the treatment of empirical "evidence" as mere material to serve the mechanist explanations of modern science. We have noted this in our previous books where we pointed to Hume's radical empiricism and how this affected Kant's notion of the empirical basis of science. So the recovery of the empirical spirit was a double edged sword. It wrecked the place of nature in the consideration of the material cause and fortified the demolition of the notion of science itself so that, with the rejection of metaphysical principles, it threw human understanding down to a raw sense level and into the pit of radical scepticism about any truth or certainty to be found, even after laborious investigations by the whole scientific community.

So it is that modern science is a mess of pseudo elements of a definition of science in terms of the four causes. But when these causal principles are taken individually some value can be seen in each of these causal approaches, if subordinated and coordinated as Aristotle did. If properly understood, not neglecting the natural forms under which they should be interpreted, material causes and empirical evidence have real value. This way we can explain how Pope Pius XII sustained an admiration not only for science in general (where he is obviously thinking of science as understood by Aristotle and Aquinas) but also for the focus on the empirical or observational side of science that has been restored in modern times. This can be seen to affect his use of the word "philosophy" in relation to science. For, as we have noted, philosophy as opposed to science can be taken for Metaphysics. But it can also be taken for the philosophical side of natural science that we have seen how Natural Philosophy is taken even in Aristotelian terms. Here we also have a use of the word "philosophical" as transcending the merely empirico-material level of natural science. Pope Pius XII at times uses the notion of philosophy, noting the lack of its being taken into account in modern thinking about science. So some care is needed in interpreting the use of the word philosophy in his speeches.

However, modern science necessarily leaves out this aspect of science when it supplants all notions of science to do with nature or the physical universe with the mixed science of physico-mathematics (dominant since Newton). Mathematical forms supplant any notion of natural forms (such as life in bodies) As we have noted, the dismissal of natural material forms also affects the way material causes are understood, emptying them as we have illustrated what happens in the arts of painting and music. The human creative imagination then plays with shapes, colours and sounds at will. Taking modern so called abstract art there are no bounds to how lines and shapes may be combined. When Chesterton was shown a modern art work supposed to be representative of a cow he remarked it looked more like a scene of San Francisco after the earthquake. Ignorant of what is going on, the art critics and public stand in admiration of this new vision of creation, a weird world of someone's imagination - the poor artist often bordering on insanity. It is similar to what happens in the world of modern science where observational and other empirical details are used as simply raw material for the creative imagination of mathematical geniuses. Sadly, the Vatican's Pontifical Academy of Science is taken in by this legerdemain. Pope Pius XII tried hard to bring them back to their senses. But other popes seem not to have been able to make the proper distinctions.

Modern science therefore has a strong hold even on the religious mind. All are subject to a universal compulsory State sponsored education system that mandates that the notion of science be used according to the modern "definition". Fancy searching for a "Theory of Everything" in Physics or Mathematics! That tells us how much the modern mind is limited to a vision of reality that is materialistic and necessarily therefore atheistic. And the shepherds do not notice that the sheep are being dispersed! It is the modern physico-mathematicians who are the real "Creationists".

We have given a general picture of modern science. Most modern scientists fit the picture, immature intellectually unable to rise about the level of mathematical abstraction in their thinking, everything is thought to be explicable in a magic mathematical formula. The created world is physical only, for which the modern scientist searches in his imagination to create a 'theory of everything'. Then endless efforts are made to "prove" modern science's universality. Novelty is sought for its own sake. We should say that there are some genuine scientists who work without paying any attention to the official description of science promulgated (perhaps ignorantly or unwittingly) by modern "philosophers of science", like Karl Popper.

The only contact with reality is the re-application of Aristotle's requirement that physical science should be verified empirically or, as Aquinas put it, terminate in the evidence of external senses or "observation".

As we have explained, this is something in modern science that was retrieved after a long period of absence in natural science after Aristotle, till towards the end of the era of Christendom. However, we have explained how this recovery, though valuable, was much less so because of the rejection of the other lines of explanation required for a complete natural science, or, in what eventuated, for a proper understanding of the mixed science of physico-mathematics, and its subordinate place in the full complex of sciences theoretical and practical as outlined by us.

Nonetheless, there was something positive recovered in the beginning of the modern era. We should keep in mind that a person is generally right in what he asserts. It is what at the same time he denies that his error consists. That was the case here. The recovery of the empirical component of natural science (or rather physico-mathematical science) was of a very low grade kind of empirical knowledge, staying at the lowest mechanical level of the understanding of material reality and bringing the same sort of mechanical level of explanation into the study of higher levels of physical reality, such as life and sense consciousness, not to speak of degrading the understanding of the spiritual and moral level of human life, it was knowledge that was useful especially when applied to human life and existence as physical or animal, but only to understanding the mechanics of biology (and sex) which modern men (and women) could then use to their advantage mainly in safe sex (which means immoral and unnatural sexual behaviour).

The success in this regard is what the modern scientist exults in and uses to ridicule and reject the philosophical and scientific achievements of the past, especially from Aristotle and Aquinas. Modern science has nothing to contribute to human life and happiness at the highest level, of spiritual insight and moral understanding which was even had in the pagan world of Plato and Aristotle. The modern mind knows its emptiness in this regard which is the reason some academics turn back to them. The modern scientist feels the poverty of his own knowledge of anything outside his specialty, knowing virtually nothing of general history except what he is fed by the official version of how modern science is an infinite improvement on all previous knowledge before it, especially when compared with the "dark ages" of religion and the era of Christendom. The modern scientist is indoctrinated in the Capitalist West as much as he was in the Communist Soviet Union. It is remarkable how the reading of the history of science is so similar in both ideologies. Both promote atheism.

Having cultivated a contempt of religion and Christianity, especially as represented in the Catholic Church, the modern scientist ridicules such giants of natural and supernatural wisdom as Saint Thomas Aquinas (and a thousand others) and spends most of his (and her) time looking at pre-modern science through dark glasses and modern science through rose coloured ones, both supplied by the State education system, whether it be Socialist or Capitalist. So ingrained is this compulsory education in "Science and Technology" that the modern youth has no chance of thinking beyond this powerful indoctrination. He gets a job within the "industrial" system if he does not buck it and retires to a life of mental vacancy if he is lucky.

There is much that can be detailed of the evils that attend this state of life in modernity. We will deal with two aspects that bring out the more pervasive and evil character of it.

There are two features among many that can be pointed to particularly with regard to modern science. They are brought to light immediately in the speeches of Pope Pius XII dealt with in the book of Frits Albers. The first is the failure to take account of the existence and providence of God as the first efficient cause of being (Maker/Alpha) and as the ultimate final cause (End/Omega) in the study of creation.

The modern scientist imagines that the only reality is the physical universe and that the things of nature are reducible to brute force (energy) and mechanical attraction (gravity) between bodies. These can be known only in terms of investigations by elemental empirical observations and mathematical/ quantitative measurements. He believes he is dealing with a world that is knowable only by the mixed science of mathematical physics, and that notion of science distorted in the manner we have described. The application of such a feature to human nature as having mind as well as body is as intellectually immature as the mathematical mind of Descartes. This pseudo spiritual aspect of his philosophy has gone nowhere. It has been given a mathematical twist and thus subtly melted into the twofold nature of modern mathematical physics. Modern philosophers could argue about the mind/body problem if they wanted to but it really became a red herring.

Pope Pius XII referred to the necessary feature of science as dependent upon holding to God as Creator. It is a requirement for one's science to be true science and we can see this from reason itself, in as much as all human understanding lacks full intelligibility if not based on appreciating that creation is a work of divine art. Just as works of human art take their forms and ends from human reason and will, so do created things and the natural order of them take their forms and ends from divine reason and will.

Lacking any notion of the primary principles of Metaphysics the modern mind cannot see the evidence for the supremacy of God both as to the existence of things and the natural order according to which they operate. This notion of Natural Law is but the notion of Divine Providence as it rules all creation. We will come to the special meaning of Natural Law as it rules human nature and its operations which, being free, are ruled by Moral Law. There we will come to the science of Ethics, which modern science explicitly excludes from the sciences Aristotle classified as practical, not theoretical.

Here we are concerned with the science of Metaphysics. This feature of true science can be seen dealt with by Aristotle in the last books of his Metaphysics (that part called Natural Theology by St. Thomas Aquinas). There is scriptural backing for this in St. Paul which the pope refers to: (cf. *Rom. 1: 19-20*)

That is the first failing into which the understanding of modern science falls, involving evident falsity, the fundamental cause being the explicit rejection of Metaphysics. This results in impotence to accept the proofs for the existence of God of Aquinas. Most Catholics today, after subjection to compulsory education in modern science, fall into this erroneous way of thinking. They are persuaded that the first proof, from motion, of St, Thomas (declared by him to be most manifest) is not valid. This they publicly proclaim even after a supposed Catholic education and calling oneself a disciple of St. Thomas (check out a recent youtube video of Matthew Fradd in his "Pints with Aquinas"). I have dealt with the proof and its metaphysical basis in

Appendix A to my book "Teenager Science and the Curse of the Creative Imagination". There are other prominent Catholic authors who imagine they can find a scientific (rational) proof for the existence of God and divine providence from a notion of science in which Mathematics has supplanted Metaphysics. It is as futile as looking for a theory of everything in mathematical physics. They do not understand that such weak efforts risk fortifying non-Catholic scientists in their atheism.

The other feature of modern science that we wish to bring out is also to do with what the modern world rejected when it revolted against the divine authority of the Catholic Church at the beginning of the modern era. It has to do with loss of any rational hold on natural morality. Though this can be discussed in general terms, we want to concentrate on that part of Moral Philosophy or Ethical Science where modern science has contributed to the most obvious devastating change in the understanding and science of human behaviour, with consequent descent of many to a level of life and conduct that as Aristotle predicted is worse than that of the most brutish animals.

We are referring to the sexual theories and practices of the present times compared with the past, even taking into account pagan and uncivilized primitive societies. Modern day sexual immorality is of such a degrading and disgraceful dimension that already one hundred years ago it prompted GKC to describe it as infinitely more depraved than that of the pagan beliefs and practices that preceded Christendom. Though it was a descent from the height of supernatural grace it could be accounted for by the moral truth that the corruption of the best is the worst.

This practical principle can be applied generally to all aspects of human life from the personal to the civil or political (even spilling over as it has in more recent times into the religious and Catholic). But our focus here is on the breakdown in understanding of and obedience to the natural moral law as it applies to the order of domestic affairs, at the heart of which are the natural moral institutions of marriage and the family.

Obviously, this subject matter cannot be discussed properly without a full understanding of Domestic Ethics. We have already noted in our book "Teenager Science and the Curse of the Creative Imagination" that we have not specially treated this part of Ethical Science. We would want to do this before providing a complete commentary on the modern situation. This we must leave to another book. Here we will only make some brief comments pointing to the main problem areas.

The Natural Moral Law is the practical equivalent to the natural theoretical principles illuminated by the Agent Intellect. This latter is a participation in the divine light of understanding that is God. It is of supreme unity (and can be attributed to the three divine persons). However, it is participated by the human intellect in such a way that produces a logical order that St. Thomas explains.

This brings in the three acts and products of the human mind, apprehension and concepts/ideas, judgment and statements/propositions and reasoning arguments and proofs. The first concept of the first act is being (ens). The first principle of the first act of judgment is the principle of non-contradiction. However, there are subsidiary concepts and ideas, and secondary principles that provide the basis for science (defined by Aristotle as "certain knowledge through causes"). So far we are talking about theoretical knowledge, and the three orders of theoretical science, Physics, Mathematics and Metaphysics (with the addition of the mixed physico-mathematical order of science discussed.

As St. Thomas explains, and we have set out in our books on the relevant subject areas, there is a parallel order in the practical/productive sciences (complemented however by arts and prudences). The first idea in this order is good. That is the transcendental notion that is the same in reality as being but distinct by a logical or mental relation (being or thing relative to inclination or desire). The first act of judgment of the human intellect as practical can be expressed in a number of ways but comes down to saying: "Seek good and shun evil". That does not tell you what is good or evil, but these are known from natural inclinations and desires following knowledge of things of nature.

We are here particularly concerned with human nature and the inclination that is free will, for that is where the natural moral order comes in. Under the very first practical principle therefore come the various primary principles of the Natural Moral Law. This by the way is what the New Natural Law

School misses altogether. So fundamentally bad is their error in this regard that one suspects that their study of Aristotle's and Aquinas's Ethics and Law has been accommodated to an understanding of modern positivist philosophy in which they are really grounded.

When we come to the evaluation of human understanding and conduct in the order of Domestic Ethics there is a huge gap in modern thinking in regard to sexual ethics. For we need to understand the important difference between primary and secondary practical principles or laws relative to the sexual union of man and woman. In common with other higher animals man (meaning to include woman) is so constituted naturally to be ordered to the reproduction of the species but then in a manner specified to the requirements of a rational animal (of spiritual soul).

Though such a natural inclination to reproduction (called procreation in man) is not the highest aspect of the union it is the most fundamental - for reproduction is the defining function of all living things, and hence of animals and man as living. Everything, as St. Thomas explains, has a primary inclination to produce its like. Indeed, it is a natural obligation.

Being rooted in the spiritual and moral level of human life, modern science is incapable of taking it into account. Modern science, even of human nature, as is clear, does not rise above the lowest mechanical (theoretical physico-mathematical) level of scientific investigations. The modern scientist, even of the highest medical and psychological expertise cannot cope with the moral character of the order of sexual acts to reproduction, let alone with the rational necessity in man to be ruled by the natural institution of marriage.

We do not go further here into domestic ethics, or any moral considerations. All we wish to point out is the abject inability of modern science to deal with these matters critically important to all aspects of human life. Thus, we see how under modern conditions of science all aspects of human life and conduct from personal to civil or political are severely and adversely impacted by this deficiency in the very essence of modern science and the scientific method applied to human sexual and family relations.

There is no need to do any "research" to verify the depth of disordered life and depraved conduct of modern life, especially experienced in "post" Christian "civilization". Nor has one to wonder where the depth of depression and extent of domestic violence have come from. Modern science can only supply mechanical type remedies such as drugs to try and alleviate the symptoms in a never ending cycle of treatment and "therapy". No one, not even Catholic scientists, tell them that the only true remedy is a return to understanding of and obedience to the Natural Moral Law. The Catholic legal philosophers and moral theologians think they can "improve on" St. Thomas Aquinas with a New Natural Law or a New Theology. The hierarchy have been "educated" in the same compulsory State education system are in no better case, even up to the highest level. They too are generally fooled by their indoctrination in the wonderful benefits of modern medical science which has freed modern man and woman from the restraints of natural morality.

We do not go into St. Thomas's explanation that the obligation to reproduce one's species is imposed on man as a species and therefore even naturally is not imposed on all men and women as individuals who of course can more directly serve higher purposes not just natural but supernatural.

What is critical to note here is that the mode of sexual union natural to the human species has a special character that differentiates it from lower animals. That is spelled out in the institution of marriage, which is natural not as animal but rational (and scripturally confirmed). It requires that the sexual union be not only between that of male and female, but also that it be confined to one man and woman to the exclusion of others for life. That is the definition of marriage as natural and mandated by the Natural Moral Law. But it is important to understand that the naturalness of this institution is not on the same level as that of the sexual union being between man and woman. The latter is a primary principle of the Natural Moral Law. Marriage follows by rational necessity from the conditions human nature imposes on what is required for the proper nurture and human education of the offspring, also from human nature as a spiritual as well as a bodily union of the spouses. We will not go further into this here.

What we wish to bring out and what is obvious is that the modern understanding and "science" of human sexual relations is gravely deficient, and essentially so. The natural moral obligation in regard to the institution of marriage is the more weakly held, as is evident from the history of mankind, even when we think of the chosen people of God. It has been obeyed fully only in the teaching and under the authority of Christ in the Catholic Church. This was quickly lost hold of after the Protestant Reformation. We do not need to detail the downward course of thinking on the nature of human sexuality that has reached its nadir in the present times.

Let that suffice to show two of the main features of the false character of modern science that is "sold" to the modern mind. Yet virtually all today including educated Catholics are blind to the faults of the modern "scientific" approach and cling onto belief in the superiority of modernity over all.

www.ingramcontent.com/pod-product-compliance
Lightning Source LLC
Chambersburg PA
CBHW070044070426
42449CB00012BA/3162